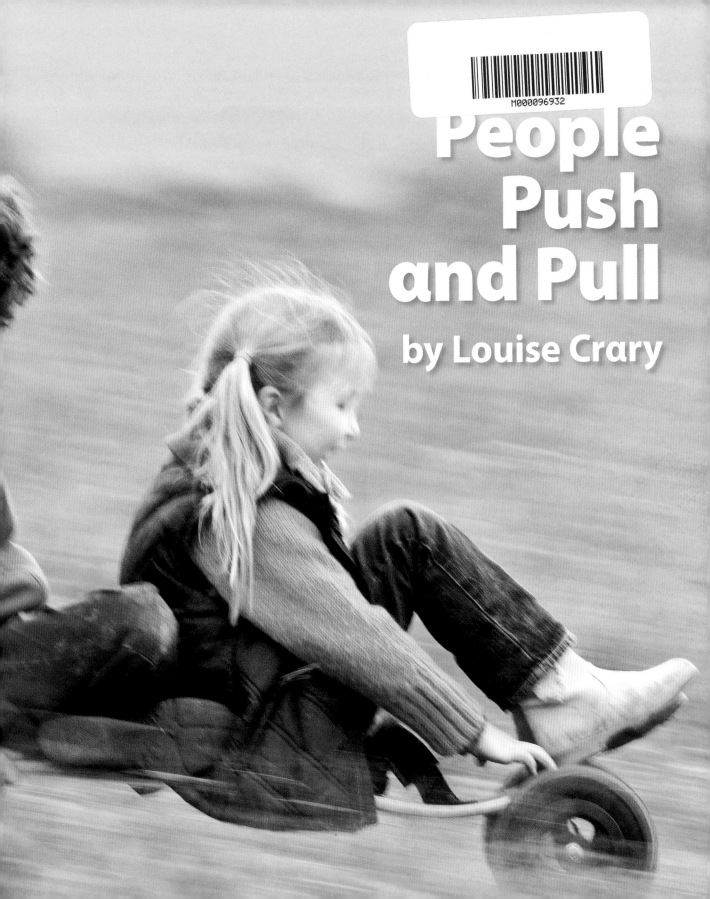

People
Push
and Pull

by Louise Crary

People **push** carts.

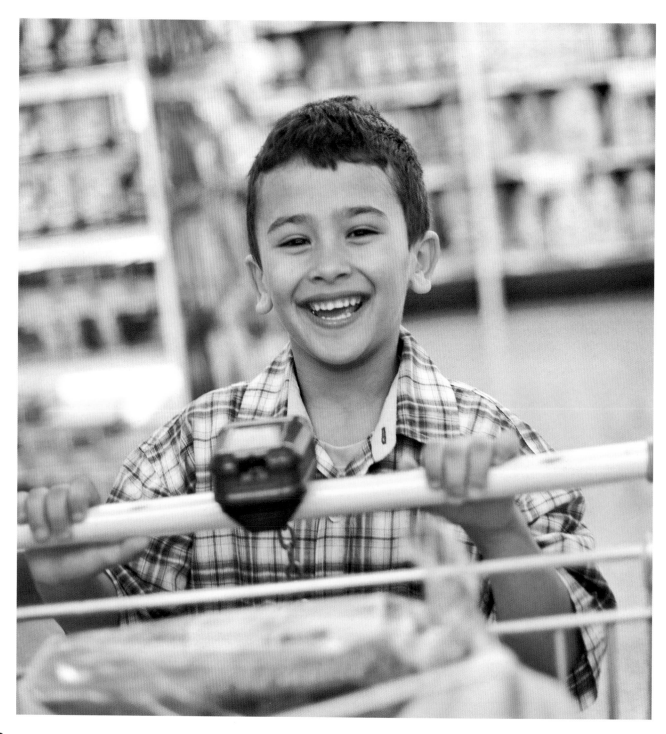

People **pull** carts.

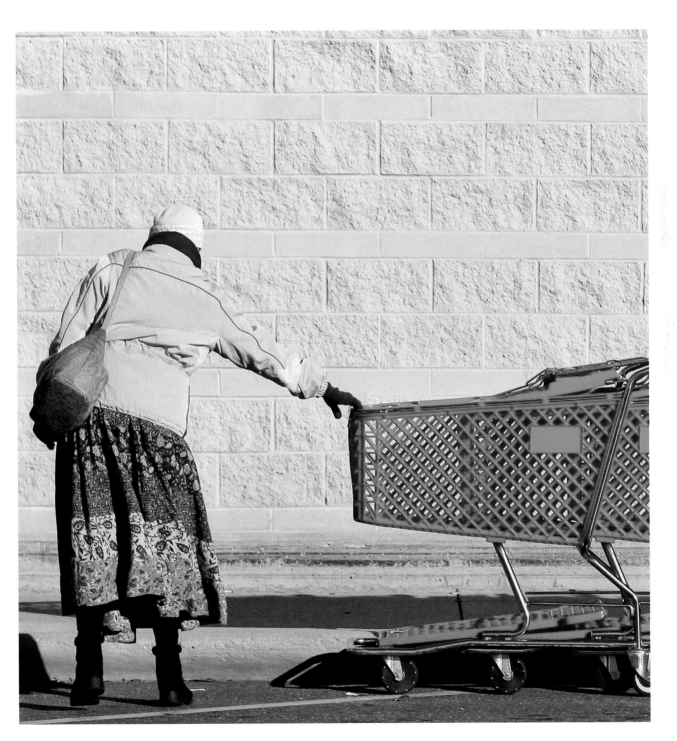

People push sleds.

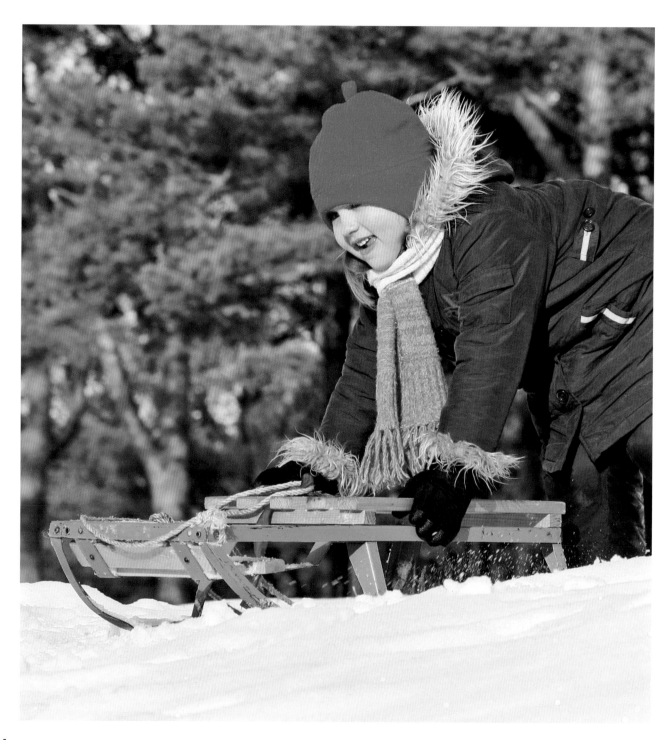

People pull sleds.

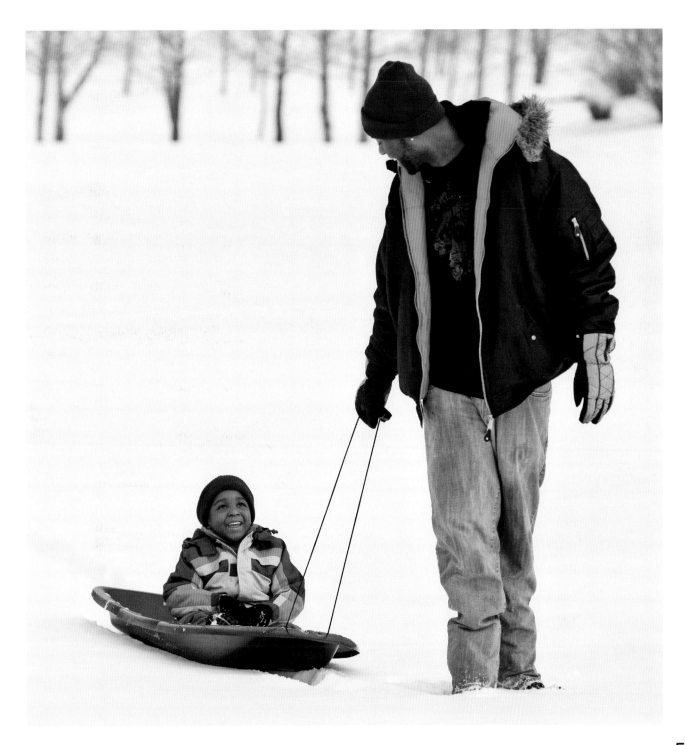

People push swings.

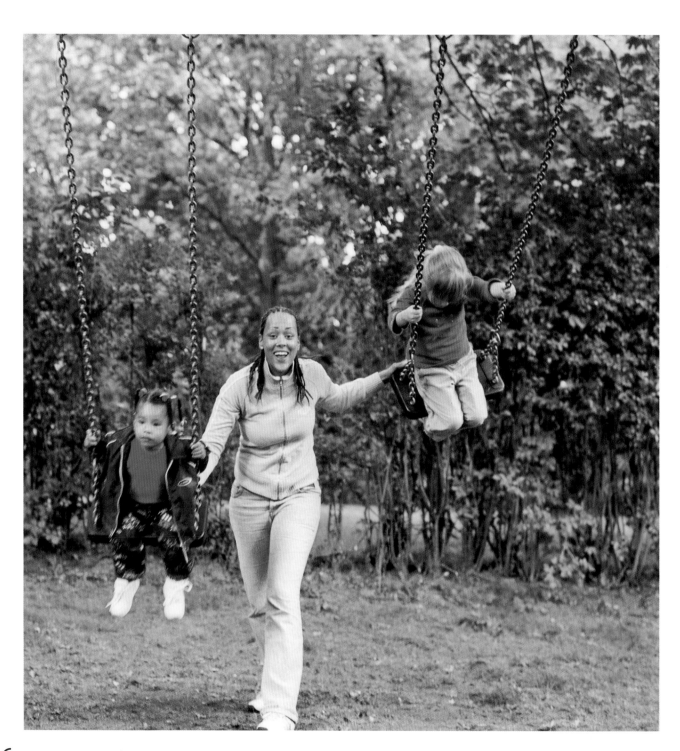

People pull swings.

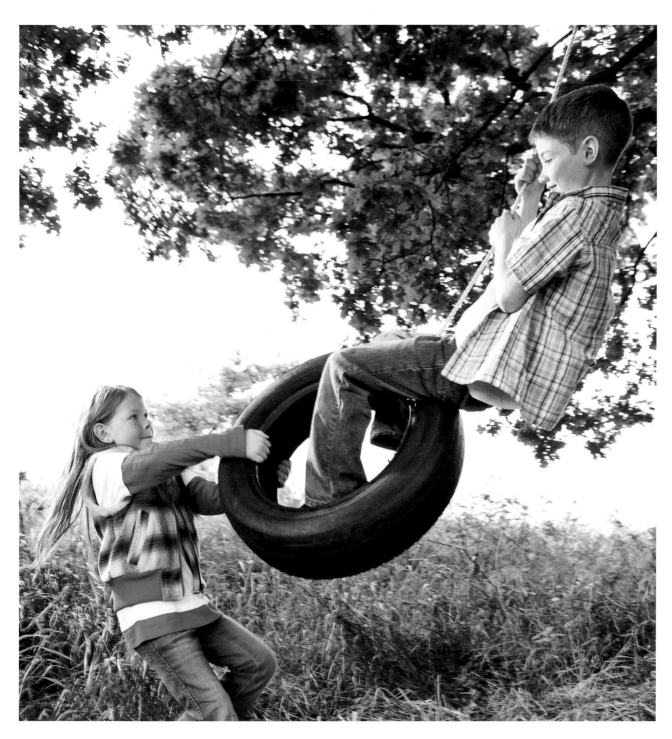

People Push and Pull

Push

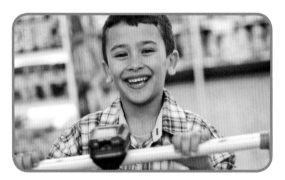

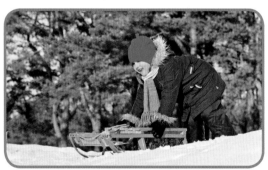

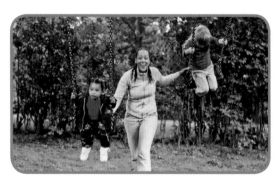

Pull

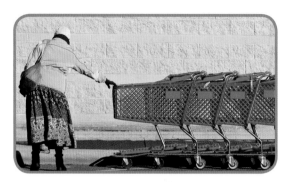

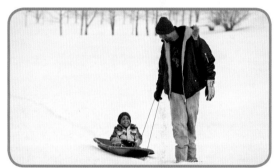

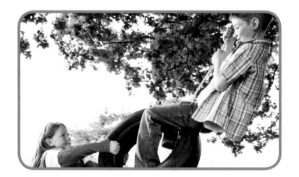